The Math Student's Companion

For All Upper Primary School Aged Students

Wilkeisha S. Bevans

ISBN: 978-1-7166-3876-3 (sc)
978-1-7163-7384-8 (e)

Library of Congress Control Number: 2020916444

Lulu Publishing Services rev. date: 04/30/2021

Table of Contents

Introduction

The Bahamas' Ministry of Education describes mathematics as, a formal discipline, which exposes hidden patterns that help us understand the world around us.

Henceforth, this booklet is designed specifically for students in the primary school system; both public and private. It is a unique easy to read and memorize mathematics study handbook. It entails the basic skills covered in the primary school Mathematics Curriculum. Parents, teachers, and students can effectively use this book alike. It is an indispensable companion for students in class and for self-testing before national and end of term examinations.

MULTIPLICATION TABLES
1-6

1 Times Table	2 Times Table	3 Times Table
1×0=0	2×0=0	3×0=0
1×1=1	2×1=2	3×1=3
1×2=2	2×2=4	3×2=6
1×3=3	2×3=6	3×3=9
1×4=4	2×4=8	3×4=12
1×5=5	2×5=10	3×5=15
1×6=6	2×6=12	3×6=18
1×7=7	2×7=14	3×7=21
1×8=8	2×8=16	3×8=24
1×9=9	2×9=18	3×9=27
1×10=10	2×10=20	3×10=30
1×11=11	2×11=22	3×11=33
1×12=12	2×12=24	3×12=36

4 Times Table	5 Times Table	6 Times Table
4×0=0	5×0=0	6×0=0
4×1=4	5×1=5	6×1=6
4×2=8	5×2=10	6×2=12
4×3=12	5×3=15	6×3=18
4×4=16	5×4=20	6×4=24
4×5=20	5×5=25	6×5=30
4×6=24	5×6=30	6×6=36
4×7=28	5×7=35	6×7=42
4×8=32	5×8=40	6×8=48
4×9=36	5×9=45	6×9=54
4×10=40	5×10=50	6×10=60
4×11=44	5×11=55	6×11=66
4×12=48	5×12=60	6×12=72

MULTIPLICATION TABLES
7-12

7 Times Table	8 Times Table	9 Times Table
7×0=0	8×0=0	9×0=0
7×1=7	8×1=8	9×1=9
7×2=14	8×2=16	9×2=18
7×3=21	8×3=24	9×3=27
7×4=28	8×4=32	9×4=36
7×5=35	8×5=40	9×5=45
7×6=42	8×6=48	9×6=54
7×7=49	8×7=56	9×7=63
7×8=56	8×8=64	9×8=72
7×9=63	8×9=72	9×9=81
7×10=70	8×10=80	9×10=90
7×11=77	8×11=88	9×11=99
7×12=84	8×12=96	9×12=108

10 Times Table	11 Times Table	12 Times Table
10×0=0	11×0=0	12×0=0
10×1=10	11×1=11	12×1=12
10×2=20	11×2=22	12×2=24
10×3=30	11×3=33	12×3=36
10×4=40	11×4=44	12×4=48
10×5=50	11×5=55	12×5=60
10×6=60	11×6=66	12×6=72
10×7=70	11×7=77	12×7=84
10×8=80	11×8=88	12×8=96
10×9=90	11×9=99	12×9=108
10×10=100	11×10=110	12×10=120
10×11=110	11×11=121	12×11=132
10×12=120	11×12=132	12×12=144

DIVISION TABLES
1-6

Division Facts 1	Division Facts 2	Division Facts 3
0÷1=0	0÷2=0	0÷3=0
1÷1=1	2÷2=1	3÷3=1
2÷1=2	4÷2=2	6÷3=2
3÷1=3	6÷2=3	9÷3=3
4÷1=4	8÷2=4	12÷3=4
5÷1=5	10÷2=5	15÷3=5
6÷1=6	12÷2=6	18÷3=6
7÷1=7	14÷2=7	21÷3=7
8÷1=8	16÷2=8	24÷3=8
8÷1=9	18÷2=9	27÷3=9
10÷1=10	20÷2=10	30÷3=10
11÷1=11	22÷2=11	33÷3=11
12÷1=12	24÷2=12	36÷3=12

Division Facts 4	Division Facts 5	Division Facts 6
0÷4=0	0÷5=0	0÷6=0
4÷4=1	5÷5=1	6÷6=1
8÷4 =2	10÷5=2	12÷6=2
12÷4=3	15÷5=3	18÷6=3
16÷4=4	20÷5=4	24÷6=4
20÷4=5	25÷5=5	30÷6=5
24÷4=6	30÷5=6	36÷6=6
28÷4=7	35÷5=7	42÷6=7
32÷4=8	40÷5=8	48÷6=8
36÷4=9	45÷5=9	54÷6=9
40÷4=10	50÷5=10	60÷6=10
44÷4=11	55÷5=11	66÷6=11
48÷4=12	60÷5=12	72÷6=12

DIVISION TABLES
7-12

Division Facts 7	Division Facts 8	Division Facts 9
0÷7=0	0÷8=0	0÷9=0
7÷7=1	8÷8=1	9÷9=1
14÷7=2	16÷8=2	18÷9=2
21÷7=3	24÷8=3	27÷9=3
28÷7=4	32÷8=4	36÷9=4
35÷7=5	40÷8=5	45÷9=5
42÷7=6	48÷8=6	54÷9=6
49÷7=7	56÷8=7	63÷9=7
56÷7=8	64÷8=8	72÷9=8
63÷7=9	72÷8=9	81÷9=9
70÷7=10	80÷8=10	90÷9=10
77÷7=11	88÷8=11	99÷9 =11
84÷7=12	96÷8=12	108÷9=12

Division Facts10	Division Facts11	Division Facts 12
0÷10=0	0÷11=0	0÷12=0
10÷10=1	11÷11=1	12÷12=1
20÷10=2	22÷11=2	24÷12=2
30÷10=3	33÷11=3	36÷12=3
40÷10=4	44÷11=4	48÷12=4
50÷10=5	55÷11=5	60÷12=5
60÷10=6	66÷11=6	72÷12=6
70÷10=7	77÷11=7	84÷12=7
80÷10=8	88÷11=8	96÷12=8
90÷10=9	99÷11=9	108÷12=9
100÷10=10	110÷11=10	120÷12=10
110÷10=11	121÷11=11	132÷12=11
120÷10=12	132÷11=12	144÷12=12

Types of Numbers

Number Types	Examples	Description
Natural Numbers	1, 2, 3, 4, 5, 6, 7, 8, 9, etc.	Numbers used for counting.
Even Numbers	2, 4, 6, 8, 10, 12, 14, 16, etc.	Numbers exactly divisible by two.
Odd Numbers	1, 3, 5, 7, 9, 11, 13, 15, etc.	Numbers not exactly divisible by two.
Prime Numbers	2, 3, 5, 7, 11, 13, 17, etc.	If a number has exactly two factors, 1 and the number itself.
Composite Numbers	4, 6, 8, 9, 10, 12, 14, etc.	If a number has more than two factors.
Whole Numbers	0, 1, 2, 3, 4, 5, 6, 7, 8, 9, etc.	Natural Numbers together with zero.
Integers	-3, -2, -1, 0, +1, +2, +3, etc.	Zero, the natural numbers and the negative naturals.
Multiples	27 is a multiple of 9 because $9 \times 3 = 27$	The product of two or more factors.
Factors	8 is a factor of 32 because 8 divides evenly into 32. $32 \div 8 = 4$	A whole number that can be divided exactly into another number.
Ordinal Numbers	1st, 2nd, 3rd, 4th, 5th, 6th, etc.	A number that indicates position.

Factors List 1 - 35

Numbers	Factors	Type
1	1	Neither
2	1,2	Prime
3	1,3	Prime
4	1,2,4	Composite
5	1,5	Prime
6	1,2,3,6	Composite
7	1,7	Prime
8	1,2,4,8	Composite
9	1,3,9	Composite
10	1,2,5,10	Composite
11	1,11	Prime
12	1,2,3,4,6,12	Composite
13	1,13	Prime
14	1,2,7,14	Composite
15	1,3,5,15	Composite
16	1,2,4,8,16	Composite
17	1,17	Prime
18	1,2,3,6,9,18	Composite
19	1,19	Prime
20	1,2,4,5,10,20	Composite
21	1,3,7,21	Composite
22	1,2,11,22	Composite
23	1,23	Prime
24	1,2,3,4,6,8,12,24	Composite
25	1,5,25	Composite
26	1,2,13,26	Composite
27	1,3,9,27	Composite
28	1,2,4,7,14,28	Composite
29	1,29	Prime
30	1,2,3,5,6,10,15,30	Composite
31	1,31	Prime
32	1,2,4,8,16,32	Composite
33	1,3,11,33	Composite
34	1,2,17,34	Composite
35	1,5,7,35	Composite

Factors List 36 - 70

Numbers	Factors	Type
36	1,2,3,4,6,8,12,18,36	Composite
37	1,37	Prime
38	1,2,19,38	Composite
39	1,3,13,39	Composite
40	1,2,4,5,8,10,20,40	Composite
41	1,41	Prime
42	1,2,2,6,7,14,21,42	Composite
43	1,43	Prime
44	1,2,4,11,22,44	Composite
45	1,3,5,9,15,45	Composite
46	1,2,23,46	Composite
47	1,47	Prime
48	1,2,3,4,6,8,12,16,24,48	Composite
49	1,7,49	Composite
50	1,2,5,10,25,50	Composite
51	1,3,17,51	Composite
52	1,2,4,13,26,52	Composite
53	1,53	Prime
54	1,2,27,54	Composite
55	1,5,11,55	Composite
56	1,2,4,7,8,14,28,56	Composite
57	1,3,19,57	Composite
58	1,2,29,58	Composite
59	1,59	Prime
60	1,2,3,4,5,6,10,12,15,20,30,60	Composite
61	1,61	Prime
62	1,2,31,62	Composite
63	1,3,7,9,21,63	Composite
64	1,2,4,8,16,32,64	Composite
65	1,5,13,65	Composite
66	1,2,3,6,11,22,33,66	Composite
67	1,67	Prime
68	1,2,4,17,34,68	Composite
69	1,3,23,69	Composite
70	1,2,5,7,10,14,35,70	Composite

Factors List 71 – 105

Numbers	Factors	Type
71	1,71	Prime
72	1, 2, 3, 4, 6, 8, 9, 12, 18, 24, 36,72	Composite
73	1,73	Prime
74	1,2,37,74	Composite
75	1,3,5,15,25,75	Composite
76	1,2,4,19,38,76	Composite
77	1,7,11,77	Composite
78	1,2,3,6,13,26,39,78	Composite
79	1,79	Prime
80	1, 2, 4, 5, 8, 10, 16, 20, 40, 80	Composite
81	1,3,9,27,81	Composite
82	1,2,41,82	Composite
83	1,83	Prime
84	1, 2, 3, 4, 6, 7, 12, 14, 21, 28, 42, 84	Composite
85	1,5,17,85	Composite
86	1,2,43,86	Composite
87	1,3,29,87	Composite
88	1,2,4,8,11,22,44,88	Composite
89	1,89	Prime
90	1, 2, 3, 5, 6, 9, 10, 15, 18, 30, 45, 90	Composite
91	1,7,13,91	Composite
92	1,2,4,23,46,92	Composite
93	1,3,31,93	Composite
94	1,2,47,94	Composite
95	1,5,19,95	Composite
96	1, 2, 3, 4, 6, 8, 12, 16, 24, 32, 48, 96	Composite
97	1,97	Prime
98	1,2,7,14,49,98	Composite
99	1,3,9,11,33,99	Composite
100	1, 2, 4, 5, 10, 20, 25, 50, 100	Composite
101	1,101	Prime
102	1, 2, 3, 6, 17, 34, 51, 102	Composite
103	1,103	Prime
104	1,2,4,8,13,26,52,104	Composite
105	1,3,5,7,15,21,35,105	Composite

Mathematical Symbols

Symbol	Description
+	addition/plus
÷ or /	division
%	percent
<	less than
>	greater than
=	equal
( )	parentheses
π	Pi (3.142)
−	subtraction/minus
×	multiplication
√	square root
≤	less than or equal to
≥	greater than or equal to
≠	not equal
[ ]	brackets
°F	degrees Fahrenheit
∴	therefore
∵	because/since
x^2	squared
x^3	cubed
x^4	power
≈	approximately equal
:	ratio
°C	degrees Celsius

Foreign Currency

The currencies below are all in correspondence with the Bahamian dollar.

COUNTRY	CURRENCY	AMOUNT
The Bahamas	Bahamian Dollar	$1.00
U.S.A.	U.S. Dollar	$1.00
Canada	Canadian Dollar	$1.33
Britain	Pound Sterling	$0.59
Europe	Euro	$0.87
India	Rupee	$71.13
Australia	Australian Dollar	$1.39
Brazil	Real	$2.22
Japan	Yen	$109.27
China	Yuan	$6.75
Columbia	Peso	$3,172.50
Switzerland	Swiss Franc	$0.99
Russia	Ruble	$66.16
Hong Kong	Hong Kong Dollar	$7.85
Jamaica	Jamaican Dollar	$132.42
Trinidad	Trinidad Dollar	$6.79
Haiti	Gourde	$78.57
Dominica	Dominican Peso	$50.44
Botswana	Pula	$10.42
Bulgaria	Lev	$1.71
Cayman Islands	Cayman Dollar	$0.83
Costa Rica	Colon	$604.52
Danish	Krone	$6.53
Egypt	Egyptian Pound	$17.62
Mexico	Peso	$19.06
Poland	Zloty	$3.75
South African	Rand	$13.70

Keywords & Phrases

Symbols	Words and Phrases Used (keywords used in problem solving)
Add ➕	add, sum, plus, increase, total, both, altogether, how many, increased, together, in all, how much, perimeter
Subtract ➖	subtract, minus, less, how much more, difference, decrease, take away, deduct, fewer, exceed, remain, are not, have left, change, decreased by, less than, how many did not have, how many more
Multiply ✖	multiply, product, by, times, lots of, total, each group, area, double, multiplied, triple
Divide ➗	divide, quotient, goes into, how many times, share, distribute, separated, average, as much, divided by, half, share equally, how many in each, cut up, each group has, parts, split

Math Terms

8 + 5 = 13

addend + addend = sum

9 – 4 = 5

minuend – subtrahend = difference

6 x 3 = 18

multiplier x multiplicand = product

5 ÷ 22 = 4 r 2

divisor ÷ dividend = quotient
& remainder

Subtraction
"Take a Penny Method"

When you are subtracting with this many zeros, you can subtract one from the subtrahend and minuend. The problem is now easier to solve.

$$
\begin{array}{r}
4\ 0\ 0\ 0 \quad {\scriptstyle -1} \\
-\ 2\ 8\ 9\ 6 \quad {\scriptstyle -1} \\
\hline
\end{array}
\qquad
\begin{array}{r}
3\ 9\ 9\ 9 \\
-\ 2\ 8\ 9\ 5 \\
\hline
1\ 1\ 0\ 4
\end{array}
$$

Do you notice something cool?
NO REGROUPING!

Long Division

DIVIDE then MULTIPLY then SUBTRACT then BRING DOWN now REPEAT

$$5 \overline{)65}$$

Ask yourself – "How many groups of 5 can I get out of 65."

Step 1: Divide

How many times will 5 go into 6? $6 \div 5 = 1$

****

Step 2: Multiply

You multiply your answer from step 1 and your divisor: $1 \times 5 = 5$

****

Step 3: Subtract

Next you subtract. $6 - 5 = 1$

****

Step 4: Bring Down

Bring down the next digit in the dividend.   5

*****

Step 5: Repeat

Step 1: Divide. ... $15 \div 5 = 3$
Step 2: Multiply. ... $5 \times 3 = 15$

Answer is **13**

Bahamian Currency

Our local currency is called the Bahamian Dollar.
It has been in circulation since 1966.

	Sir Lyden Oscar Pindling (First Prime Minister)
	Queen Elizabeth II (Monarch-Head of State)
	Sir Cecil Wallace- Whitfield (Founder of The FNM)
	Sir Stafford Sands (The Father of Tourism)
	Sir Milo Butler (First Governor General)
	Sir Rolland Symonette (First Premier)
	Queen Elizabeth II (Monarch-Head of State)

Mean, Median, Mode and Range

Finding the mean, median, mode &
range for this set of data:

85%, 75% 100%, 69%, 90%, 85%, 80%.

Mean is the average of a set of numbers.
Find the sum of the data items. Divide the
sum by the number of items.

85 + 75 + 100 + 69 +90 + 85 + 80 = 584
584 ÷ 7 = 83.4

Mean = 83.4

Mode: Arrange the items from least to
greatest. The mode is the number that is
repeated most often. If no value repeats
there is no mode.

69%, 75%, 80%, 85%, 85%, 90%, 100%

Mode = 85%

Mean, Median, Mode and Range

Range: Arrange the items from least to greatest. The range is the difference of the greatest and least values.

69%, 75%, 80%, 85%, 85%, 90%, 100%

100—69 = 31
Range = 31

Median is the middle number when a set of numbers are written in order from least to greatest.

69%, 75%, 80%, 85%, 85%, 90%, 100%
Median = 85%

*If there are two middle values the median is the average of the two values.

6, 6, 9, 11, 15, 19

9 + 11 = 20 ÷ 2 = 10
Median = 10

Lines & Rays

TERMS AND DEFINITION	RAYS & LINES
A **point** names an exact location in space.	• A
A **line** is a straight path of points that continues without end in both directions. There are no end points.	B C
A **line segment** is a part of a line. It has two endpoints and all the points between them. It's the shortest distance between two points.	B C
A **ray** is part of a line. It has one endpoint and continues without end in one direction.	B C
A **plane** is a flat surface of points that continues without end in all directions.	
Parallel Lines are lines in the same plane that never intersect and are always the same distance apart.	
Intersecting Lines are lines that cross each other at exactly one point. They form four angles.	
Perpendicular Lines are lines that intersect to form four right angles.	

Place Value

BILLIONS			MILLIONS			THOUSANDS			ONES		
Hundreds	Tens	Ones	Hundreds	Tens	Ones	Hundreds	Tens	Ones	Hundreds	Tens	Ones
	1	9,	4	0	7,	1	3	4,	5	6	0

Standard Form: 19,407,134,560

Expanded Form: 10,000,000,000 + 9,000,000,000 + 400,000,000 + 7,000,000 + 100,000 + 30,000 + 4,000 + 500 + 60 + 0

Word Form: Nineteen billion, four hundred seven million, one hundred thirty-four thousand, five hundred sixty

Short Word Form: 19 billion, 407 million, 134 thousand, 560

Place Value with Decimals

	ONES	TENTHS	HUNDREDTHS	THOUSANDSTHS	TEN-THOUSANDTHS	HUNDRED-THOUSANDSTHS
	2 .	7	8	3	5	0
	2	0.7	0.08	0.003	0.0005	0

Standard Form: 2.78350

Word Form: two and seven thousand, eight hundred thirty - five ten thousandths

Expanded Form: 2 + 0.7 + 0.08 + 0.003 + 0.0005

	3	4	5	6	7	8	9	10	11	12	13	14	15	16	17	18	19	20
1	33	25	20	17	14	13	11	10	9	8	8	7	7	6	6	6	5	5
2	67	50	40	33	29	25	22	20	18	17	15	14	13	13	12	11	11	10
3	100	75	60	50	43	38	33	30	27	25	23	21	20	19	18	17	16	15
4		100	80	67	57	50	44	40	36	33	31	29	27	25	24	22	21	20
5			100	83	71	63	56	50	45	42	38	36	33	31	29	28	26	25
6				100	86	75	67	60	55	50	46	43	40	38	35	33	32	30
7					100	88	78	70	64	58	54	50	47	44	41	39	37	35
8						100	89	80	73	67	62	57	53	50	47	44	42	40
9							100	90	82	75	69	64	60	56	53	50	47	45
10								100	91	83	77	71	67	63	59	56	53	50
11									100	92	85	79	73	69	65	61	58	55
12										100	92	86	80	75	71	67	63	60
13											100	93	87	81	76	72	68	65
14												100	93	88	82	78	74	70
15													100	93	88	83	79	75
16														100	94	89	84	80
17															100	94	89	85
18																100	94	90
19																	100	95
20																		100

Directions: Look at the number running across the top of the chart. Find the total number of possible right answers in the exercise. Then find the number of answers you had right in the column at the left of the chart. Find the place where the two lines meet. This will be your percentage of correct answers.

Example:
If there are 20 points in an assignment, but the child only got 15, then find 15 along the left side. Now, move your fingers right along the chart until you find where the two lines meet. The answer is 75. Therefore 15 out of 20 gives you 75% B.

Actual Numbers Right

22

Measurement

Capacity is the amount a container can hold. This unit measures liquids.

Customary Units of Capacity
8 fluid ounces (fl oz.) =1 cup(c)
2cups=1 pint (pt.)
2 pints=1 quart (qt.)
4 cups=1 quart
4 quarts=1 gallon(gal)

Metric Units of Capacity
1,000 milliliters (mL) = 1 liter(L)
250 milliliters = 1 metric cup
4 metric cups = 1 liter
1,000 liters = 1 kiloliter (kL)

Length is measurement of distance, width or height.

Customary Units of Length
12 inches (in.) = 1 foot (ft.)
3 feet = 1 yard (yd.)
5,280 feet = 1 mile (mi)
1,760 yards = 1 mile (mi)

Metric Units of Length
10 millimeters (mm) = 1 centimeter (cm)
100 centimeters = 1 meter (m)
1,000 meters = 1 kilometer (km)

Measurement

Mass is the amount of matter in an object. It also measures the weight on an object.

Customary Units of Weight
16 ounces (oz.) = 1 pound (lb.)
2,000 pounds = 1 ton (T)

Metric Units of Weight
1,000 milligrams (mg) = 1gram (g)
1,000 grams = 1 kilogram (kg)
1kilogram = 100,000 centigrams (cg)
1 gram = 10 decigram

Time Units of Time

60 seconds (sec) = 1 minutes (min)
60 minutes = 1 hour (hr)
24 hours = 1 day
7 days = 1 week (wk)
52 weeks = 1 years (yr)

365 days = 1 year
366 days = 1 leap year
10 years = 1 decade
100 years = 1 century
1,000 years = 1 millennium

Temperature is the hotness or coldness of a body of environment. It is measured in degrees.

Fahrenheit (F)   Celsius (C)

Measurement

Conversion of Units

To change smaller units to larger units, divide.

e.g.	24 in = 2 ft.
because	12 in = 1 ft.
∴	24 ÷ 12 = 2

Metric Conversion Chart

To convert to a smaller unit, move decimal point to the right or multiply.

Kilo- 1,000 units	Hecto- 100 units	Deca- 10 units	Basic Unit	Deci- 0.1 units	Centi- 0.01 units	Milli- 0.001 units

To convert to a larger unit, move decimal point to the left or divide.

Conversion of Units

To change larger units to smaller units, multiply.

e.g.	4,000 g = 4 kg
because	1,000 = 1 kg
∴	1,000 x 4 = 4,000

Angles

An <u>angle</u> is the amount of turning between two lines that meet at a point.

Parts of an Angle

The corner point of an angle is called the vertex.
The two straight sides are called *arms*.
The angle is the *amount of turn* between each arm.

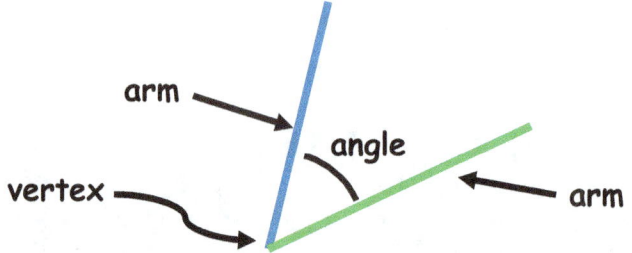

A reflex angle is more than 180° but less than 360°.

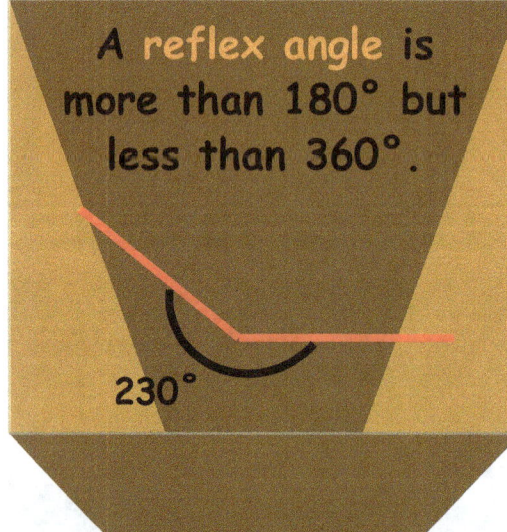

An acute angle is less than 90°.

Angles

A **straight angle** is 180°.

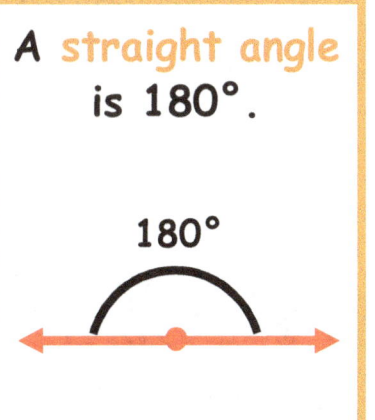

180°

A **right angle** is an internal angle which is equal to 90°.

90°

An **obtuse angle** is more than 90° but less than 180°.

130°

A **protractor** is a measuring instrument, typically made of transparent plastic or glass, and is used for measuring angles.

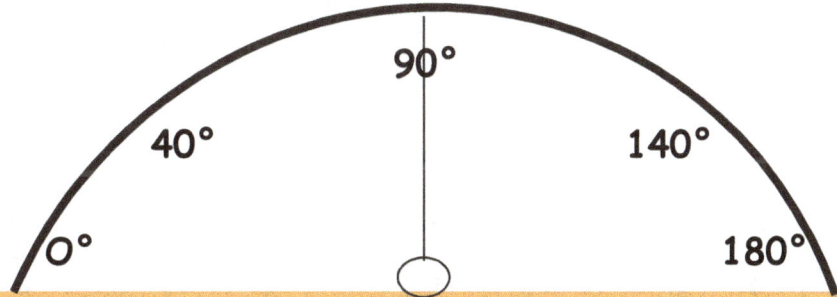

90°

40°

140°

0°

180°

Rules of Divisibility

A number is divisible by another number when the quotient is a whole number an there is a remainder of zero.

A NUMBER IS DIVISIBLE BY...	
2	if the last digit is even (*e.g. 0, 2, 4, 6 or 8*).
3	if the sum of the digits is divisible by 3.
4	if the last two digits form a number divisible by 4
5	if the last digit is 0 or 5.
6	if the number is divisible by both 2 and 3
9	if the sum of the digits is divisible by 9.
10	if the last digit is 0.

EXAMPLES		
DIVISIBLE	by	NOT DIVISIBLE
324 or 3,978	2	975
612 or 315	3	139
8,512 or 6,028	4	518
975 or 800	5	1,978
48 or 3,132	6	20
423 or 711	9	93
990 or 450	10	536

Integers

An **integer** is a number on the number line. It is not a fraction, it is a whole number.

$$…,-3, -2, -1, 0, 1, 2, 3,…$$

Notice that all of the whole numbers are also integers. The illustration below shows the integers graphed on the number line. The integers include zero and continue to the right and to the left.

Number Line

Number Lines helps us tell which numbers are greater or lesser.

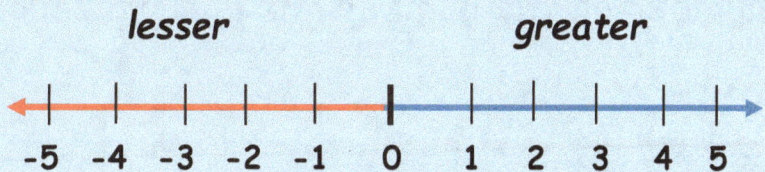

Zero is neither negative nor positive.

Negative Numbers (-) continue left nonstop.
Positive Numbers (+) continue right nonstop.

A number on the **left is less** than a number on the right.
Examples:

$$7 \text{ is less than } 12$$
$$-1 \text{ is less than } 1$$
$$-6 \text{ is less than } -3$$

A number on the **right is greater** than a number on the left.
Examples:

$$9 \text{ is greater than } 6$$
$$1 \text{ is greater than } -1$$
$$-2 \text{ is greater than } -7$$

Absolute Value

An <u>absolute</u> value is basically how far a number on the number line is from zero.

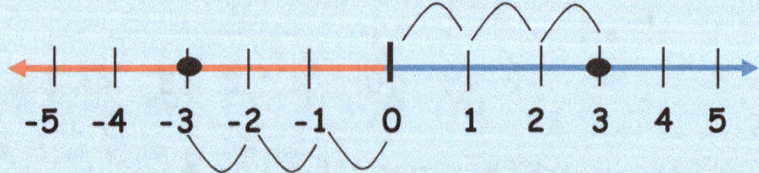

For instance: The absolute value of |3| is 3, because it's three units away from zero. The absolute value of |-3| is also 3.

Roman Numerals

Ancient Romans used seven numbers that look different from ours. We call this system Roman Numerals. Roman Numbers are used today on clocks/watches, pages of books, Sporting events i.e. The Super bowl, referring to royalty i.e. Queen Elizabeth II etc.

Roman Numerals	I	V	X	L	C	D	M
Our Numbers	1	5	10	50	100	500	1,000

RULES	EXAMPLES
When a letter is repeated, add the value of each letter.	XXX X + X + X = 30
When a letter with a greater value is placed before a smaller one, you add.	XVI 10 + 5 + 1 = 16
If a smaller value is placed before a larger one, you subtract.	XL 50 – 10 = 40
Whenever there is a 4 or 9 in Roman Numerals, you must subtract.	IX 10 – 1 = 9

MODERN NUMBERS	ROMAN NUMERALS	MODERN NUMBERS	ROMAN NUMERALS
1	I	30	XXX
2	II	35	XXXV
3	III	40	XL
4	IV	45	XLV
5	V	50	L
6	VI	55	LV
7	VII	60	LX
8	VIII	65	LXV
9	IX	70	LXX
10	X	75	LXXV
11	XI	80	LXXX
12	XII	85	LXXXV
13	XIII	90	XC
14	XIV	95	XCV
15	XV	100	C
16	XVI	200	CC
17	XVII	300	CCC
18	XVIII	400	CD
19	XIX	500	D
20	XX	600	DC
21	XXI	700	DCC
22	XXII	800	DCCC
23	XXIII	900	CM
24	XXIV	1000	M
25	XXV	2000	MM

Order of Operations

The **order of operations** is a rule that tells you the sequence to follow when you are performing operations in a magical expression.

1.	2.	3.	4.
parenthesis	exponent	multiplication	addition
P	E	M	A
		division	subtraction
		D	S
( )	y ˣ	× ÷	+ -

Do **P**, then **E**. Then do **M** or **D**, left to right. Lastly, do **A** or **S** left to right.

$$127 - 2 (3 + 4)^2$$
$$= 127 - 2 (7)^2$$
$$= 127 - 2 \times 49$$
$$= 127 - 98$$
$$= 29$$

Symmetry

A shape has _symmetry_ if both halves match exactly when it is folded on the axis of symmetry. A shape is _asymmetrical_ if it doesn't have any matches exactly when it is folded.

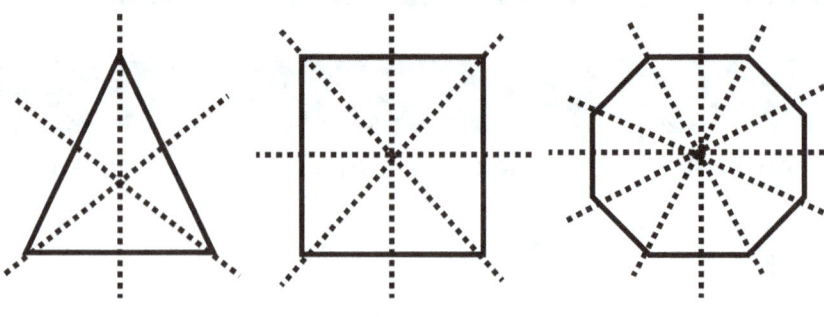

Tally

A mark made to represent an item when counting. Four vertical marks are made and crosses horizontally with the fifth to keep them in bundles of fives.

1	I	6	✚ I
2	II	7	✚ II
3	III	8	✚ III
4	IIII	9	✚ IIII
5	✚	10	✚ ✚

Problem Solving Techniques

Eight helpful techniques to use when solving math word problems.

Make a Table

Sample Question	Sample of Strategy
You save $3 on Monday. Each day after that you save twice as much as you saved the day before. If this pattern continues, how much would you save on Friday?	<table><tr><th>Days</th><th>Amount Saved</th></tr><tr><td>Monday</td><td>$3</td></tr><tr><td>Tuesday</td><td>$6</td></tr><tr><td>Wednesday</td><td>$12</td></tr><tr><td>Thursday</td><td>$24</td></tr><tr><td>Friday</td><td>$48</td></tr></table>

Find a Pattern

Sample Question	Sample of Strategy
Amare has written a number pattern that begins with 1, 3, 6, 10, and 15. If he continues this pattern, what are the next four numbers in his pattern?	3 = 1 + 2 (starting number is 1, add 2 to make 3) 6 = 3 + 3 (starting number is 3, add 3 to make 6) 10 = 6 + 4 (starting number is 6, add 4 to make 10) 15 = 10 + 5 (starting number is 10, add 5 to make 15) etc.

Work Backwards

Sample Question	Sample of Strategy
Ethan walked from South Bahamia to Queen's Hwy. It took 1 hr. 25 min. to walk from South Bahamia to Downtown. Then it took 25 min. to walk from Downtown to Queen's Hwy. He arrived on Queen's Hwy at 2:45 pm. What time did he leave South Bahamia?	Start at 2:45. This is the time Ethan reached Queen's Hwy. Subtract 25 minutes. This is the time it took to get from Downtown to Queen's Hwy. Time is: 2:20 P.M. Subtract: 1 hour 25 minutes. This is the time it took to get from South Bahamia to Downtown. Ethan left South Bahamia at 12:55 P.M.

Guess & Check

Sample Question	Sample of Strategy
Kiara and Shae sold 12 show tickets altogether. Kiara sold 2 more tickets than Shae. How many tickets did each girl sell?	**Guess:** Kiara = 7 tickets Shae = 5 tickets **Check:** 7 + 5 = 12 7 - 5 = 2 (Kiara sold 2 more tickets) These numbers work! Kiara sold 7 tickets and Shae sold 5 tickets.

Draw a Picture

Sample Question	Sample of Strategy
Damar has 3 yellow, 4 aqua and 1 black chip in his bag. What fraction are yellow chips?	⭕⭕⭕🔵🔵🔵🔵⚫ 3/8 of the chips are yellow.

Make a List

Sample Question	Sample of Strategy
Milano is taking pictures of Jay, Kassius and Angelo. He asks them, "How many different ways could you three children stand in a line?"	<table><tr><td>First</td><td>Second</td><td>Third</td></tr><tr><td>Jay</td><td>Kassius</td><td>Angelo</td></tr><tr><td>Jay</td><td>Angelo</td><td>Kassius</td></tr><tr><td>Kassius</td><td>Jay</td><td>Angelo</td></tr><tr><td>Kassius</td><td>Angelo</td><td>Jay</td></tr><tr><td>Angelo</td><td>Kassius</td><td>Jay</td></tr><tr><td>Angelo</td><td>Jay</td><td>Kassius</td></tr></table>

Write a Number Sentence

Sample Question	Sample of Strategy
Ava puts 18 pencils in 3 equal groups. How many pencils are in each group?	$18 \div 3 = 6$

Use Logical Reasoning

Sample Question	Sample of Strategy
3 birds & one bird-cage cost the same as 5 birds. If one bird costs $3.50, what does a bird-cage cost?	Use Common Sense! Write to help explain your best thinking using words, numbers, or pictures. The answer is $6

PROPERTIES

There are three basic properties of numbers;

COMMUTATIVE,
ASSOCIATIVE,
AND DISTRIBUTIVE.

The word "<u>commutative</u>" comes from "commute" or "move around".

The word "<u>associative</u>" comes from "associate" or "group"; the Associative Property is the rule that refers to grouping.

The <u>Distributive Property</u> is easy to remember, if you recall multiplication distributes over addition.

PROPERTIES OF ADDITION

Zero Property of Addition:

States that when you add 0 to any real number, the sum is the number itself.

$$9 + n = 9$$

$$9 + 0 = 9$$

Commutative/Order Property of Addition:

States that changing the order of the addends will not affect the sum. We can **swap numbers** over and still get the same answer.

$$17 + 4 = n + 17$$

$$17 + 4 = 4 + 17$$

Associative/Grouping Property of Addition:

States that changing the groupings of the addends will not affect the sum. It doesn't matter how we group the numbers.

$$(12 + 8) + 3 = 12 + (n + 3)$$

$$(12 + 8) + 3 = 12 + (8 + 3)$$

PROPERTIES OF MULTIPLICATION

Associative/Grouping Property of Multiplication:

States that changing the groupings of the factor will not affect the product.

$$(8 \times n) \times 5 = 8 \times (6 \times 5)$$

$$(8 \times 6) \times 5 = 8 \times (6 \times 5)$$

Property of One:

States that when you multiply any number by 1 the result is the number.

$$n \times 1 = 16$$

$$16 \times 1 = 16$$

Commutative Property of Multiplication:

States that changing the order of the factors will not affect the product.

$$4 \times 3 = 3 \times 4$$

$$4 \times 3 = 3 \times 4$$

Zero Property of Multiplication:

States that when you multiply any number by 0 the result is 0.

$$7 \times n = 0$$

$$7 \times 0 = 0$$

Distributive Property of Multiplication:

States that multiplying a sum by a number is the same as multiplying each added in the sum by the number then adding the products.

$$6 \times (n + 5) = (6 \times n) + (6 \times 5)$$

$$6 \times (9 + 5) = (6 \times 9) + (6 \times 5)$$

Quadrilaterals

A **quadrilateral** is a four-sided polygon with four angles. There are many kinds of quadrilaterals. The five most common types are the <u>parallelogram</u>, the <u>rectangle</u>, the <u>square</u>, the <u>trapezoid</u>, the <u>kite</u> and the <u>rhombus</u>.

Quadrilaterals	Description
Parallelogram	Opposite sides are equal. Opposite sides are parallel. Opposite angles are equal.
Rhombus	Opposite sides are parallel. All sides are equal.
Rectangle	Opposite sides are equal. Opposite sides are parallel. All angles are right angles.
Square	Opposite sides are parallel. All sides are equal. All angles are right angles.
Trapezoid	Only one pair of opposite sides are parallel.
Kite	Exactly two pairs of consecutive sides are equal.

Polygon

A polygon is a closed plane figure formed by three or more lines. Polygons are named by the number of their sides and angles. In a regular polygon, all the sides have equal lengths and all the angles have equal measures. Irregular polygons are the exact opposite.

Figure	Sides & Angles	Regular Polygon	Irregular Polygons
Triangle	3		
Quadrilateral	4		
Pentagon	5		
Hexagon	6		
Octagon	8		

Big Numbers

The largest number that has a commonly-known specific name is a "googolplex", which is a 1 followed by a googol zeros, where a "googol" is (a 1 followed by 100 zeros).

A **Googol** is 1 followed by one hundred zeros (10^{100}) :
10,000,000,000,000,000,000,000,000,000,000
000,000,000,000,000,000,000,000,000,000,
000,000,000,000,000,000,000,000,000,000.

Whole Numbers	Decimals
Million	Millionth
Billion	Billionth
Trillion	Trillionth
Quadrillion	Quadrillionth
Quintillion	Quintillionth
Sextillion	Sextillionth
Septillion	Septillionth
Octillion	Octillionth
Nonillion	Nonillionth
Decillion	Decillionth
Undecillion	Undecillionth
Duodecillion	Duodecillionth
Tredecillion	Tredecillionth
Quattuordecillion	Quattuordecillionth
Quindecillion	Quindecillionth
Sexdecillion	Sexdecillionth
Septemdecillion	Septemdecillionth
Octodecillion	Octodecillionth
Novemdecillion	Novemdecillionth
Vigintillion	Vigintillionth
Centillion	Centillionth

Day and Night

The Earth takes 24 hours to make one complete turn. We have day and night because the Earth rotates. At any moment, half of the world is in daytime and half is in nighttime.

	Rotation	Revolution
Definition	The earth rotates (spins or turns) on its own axis.	The earth revolves (orbits) around the sun.
Time	takes 24hrs or 1 day 12:00 am - 12:00 am	takes 365 days or 1 year Jan. 1st - Dec. 31st
Causes	causes day and night	The four seasons (Summer, Winter, Spring, Fall)

Midnight

Midday

Midnight

| 12:00 | 3:00 | 6:00 | 9:00 | 12:00 | 3:00 | 6:00 | 9:00 | 12:00 |

am ⟶ pm

a.m. antemeridian : midnight to noon
12:00am to 12:00pm (before noon)

p.m. postmeridian : noon to midnight
12:00pm to 12:00am (after noon)

Ordered Pairs

Ordered pairs are a pair of numbers used to locate a point on a coordinate plane.

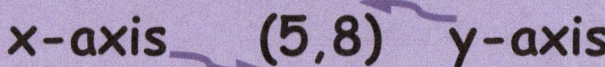

x-axis   (5,8)   y-axis

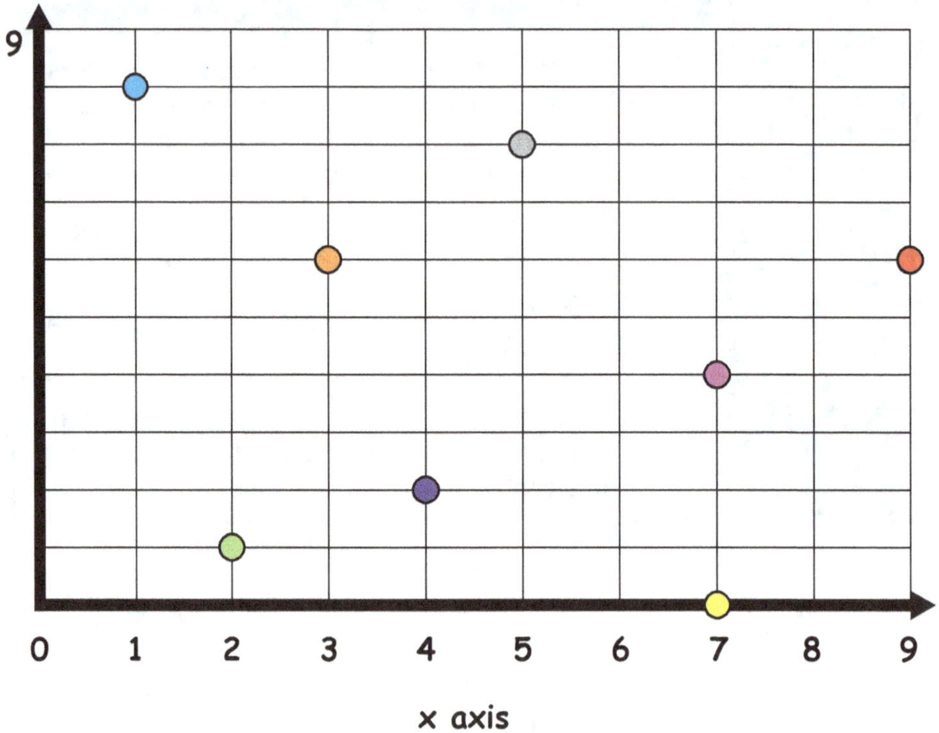

9

0   1   2   3   4   5   6   7   8   9

x axis

◯ (1,9)   ◯ (3,6)   ◯ (5,8)   ◯ (7,0)

◯ (2,1)   ◯ (4,2)   ◯ (9,6)   ◯ (8,4)

ORIGIN (0,0)

To find the coordinate:
- start at the origin
- move right across the x – axis
- then move up the y – axis.

Important Formulas

Perimeter

The perimeter is the length of the outside boundary of a shape.

Polygon: P = number of sides x sides

7mm

P = # x s
= 5 x 7mm
= <u>35mm</u>

Rectangle: P = (2 x length) + (2 x width)

10 yds.

5 yds.

P = (2 x l) + (2 x w)
= (2 x 10 yds.) + (2 x 5 yds.)
= 20 yds. + 10 yds.
= <u>30 yds.</u>

Square: P = 4 x sides

3 in.

3 in.

P = 4 x s
= 4 x 3 in
= <u>12in</u>

Irregular Polygon: P = s + s + s + s

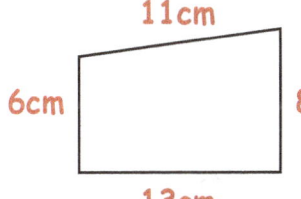

11cm

6cm   8cm

13cm

P= 11cm+8cm+6cm+13cm
= <u>38cm</u>

Important Formulas

Area

The area is the size of the surface. It measures in square units.

Area of square: $A = \text{side} \times \text{side}$

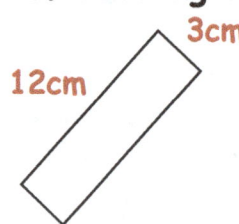

6cm

$A = s \times s$
$= 6cm \times 6cm$
$= \underline{36cm^2}$

Area of rectangle: $A = \text{length} \times \text{width}$

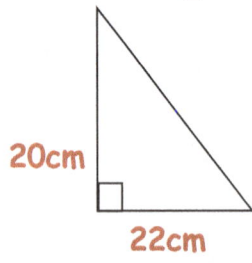

3cm

12cm

$A = l \times w$
$= 12cm \times 3cm$
$= \underline{36cm^2}$

Area of triangle: $A = \frac{1}{2} \times (\text{base} \times \text{height})$

20cm

22cm

$A = \frac{1}{2} \times (b \times h)$
$= \frac{1}{2} \times (22cm \times 20cm)$
$= \frac{1}{2} \times 440cm$
$= \underline{220cm^2}$

Area of parallelogram: $A = \text{base} \times \text{height}$

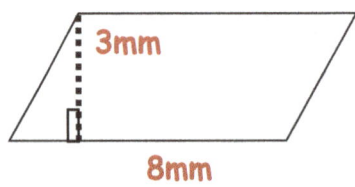

3mm

8mm

$A = b \times h$
$= 8mm \times 3mm$
$= \underline{24mm^2}$

Important Formulas

Volume

The volume is the amount of space a solid object takes up.

Volume of rectangular prism:
v = length x width x height
$\quad$ = 10cm x 6cm x 5cm
$\quad$ = 60cm x 5 cm
$\quad$ = 300cm^3

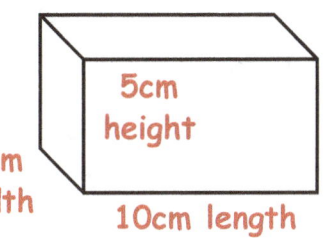

5cm height
6cm width
10cm length

Volume of a cylinder:

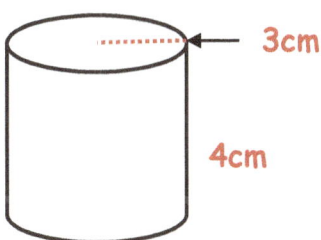

3cm
4cm

A cylinder has a circular base. So, use the formula πr^2 to find the area of the base,
A=3.142 x 3cm^2
$\quad$ =3.142 x 9cm
$\quad$ =28.278cm^2

The volume of the cylinder is equal to its base area times its height.
V=28.278cm^{2}x 4cm
$\quad$ =113.112cm^3

Circumference

The distance around a circle is called the circumference.

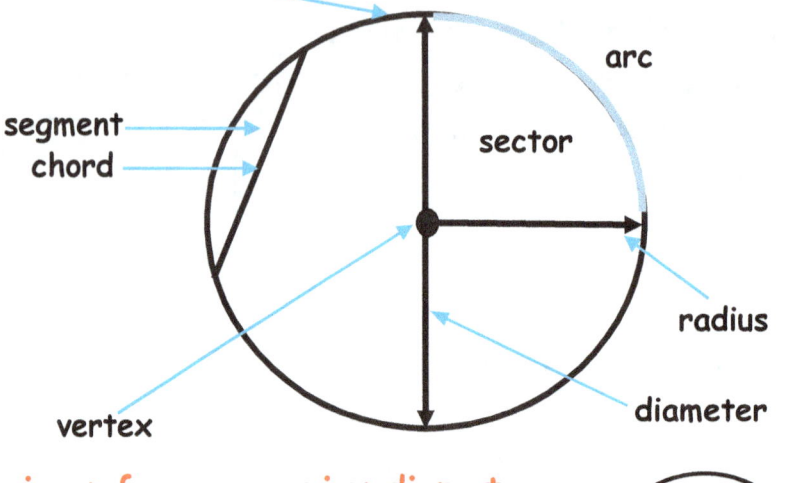

circumference = pi x diameter
= π x d
= 3.142 x 18in
= 56.556 in

The radius is half of the diameter. So, if the diameter is 18 inches, then the radius would be 8 inches.

The number π is a mathematical constant. This means that it is a very special number. Pi has trillions of digits. The first few are:

3.14159265358979323846264338327950288841971693....

Time

Time can be displayed in 12 hour (civilian time) or 24-hour (military time).

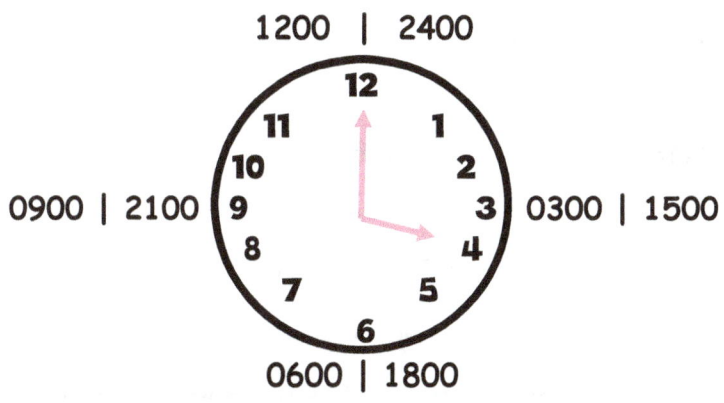

1200 | 2400

0900 | 2100

0300 | 1500

0600 | 1800

CIVILIAN	MILITARY	CIVILIAN	MILITARY
12:00 am	0000 midnight	12:00 pm	1200 noon
1:00 am	0100	1:00 pm	1300
2:00 am	0200	2:00 pm	1400
3:00 am	0300	3:00 pm	1500
4:00 am	0400	4:00 pm	1600
5:00 am	0500	5:00 pm	1700
6:00 am	0600	6:00 pm	1800
7:00 am	0700	7:00 pm	1900
8:00 am	0800	8:00 pm	2000
9:00 am	0900	9:00 pm	2100
10:00 am	1000	10:00 pm	2200
11:00 am	1100	11:00 pm	2300

ANALOG CLOCK

DIGITAL CLOCK

System International

The International System of Units is also referred to as SI Format. It is used to abbreviate dates. The units are placed in order starting from the largest unit.

YEAR	MONTH	DAY
1958	09	06

or

1958/9/6

The most commonly used abbreviated date format in The Bahamas uses the format of day, month, then year.

DAY	MONTH	YEARS
22	11	1985

or

22/11/1985

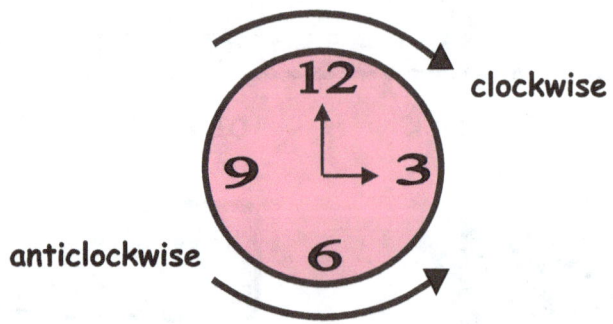

clockwise

anticlockwise

Transformation

A movement of a figure without changing the size or shape of the figure is a rigid transformation. Since the size and shape do not change, the original figure and the transformation are always congruent.

A <u>translation</u> or <u>slide</u> is the movement of a figure along a straight line. Only the location of the figure changes with translation.

Turning a figure around a point is called <u>rotation</u> or <u>turn</u>. Both the position and the location of the figure can change. A point of rotation can be on or outside a figure.

Flipping a figure over a line is called a <u>reflection</u> or <u>flip</u> about that line. Both the position and the location of the figure change with reflection.

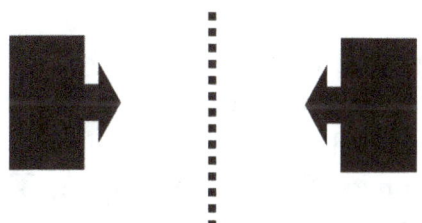

Rounding

Rounding means reducing the digits in a number while trying to keep its value similar. Leave it the same if the next digit is less than 5 (this is called *rounding down*). Increase it by one if the next digit is 5 or more (this is called *rounding up*).

Rounding Whole Numbers

Examples	Because...
3**74** rounded to the nearest tens is **370**	...the last digit (4) is less than 5 so –round down
1**37** rounded to the nearest tens is **140**	...the next digit (7) is 5 or more so – round up

Round 3**74** to the nearest ten.

* Find the rounding place, [tens]
* Ask yourself:
a. What digit do I look at? The digit next door [4].
b. Is it less or more than 5? It is less than 5.
c. What happens to the digit next door? It remains the same-7.
d. What happens to the digit to the left of the tens place? It remains the same-3.

   What happens to the digit to the right of the tens place? It becomes a zero. Therefore, **374** rounded to the nearest ten is **370**.

Rounding

Rounding Decimals

Examples	Because...
3.71 rounded to the nearest tenths is **3.70**	...the next digit (1) is less than 5 – so round down
1.2635 rounded to tenths is 1.3	...the next digit (6) is 5 or more – so round up

Hundreds Chart

Round down ⟵   Round up ⟶

0	1	2	3	4	5	6	7	8	9
10	11	12	13	14	15	16	17	18	19
20	21	22	23	24	25	26	27	28	29
30	31	32	33	34	35	36	37	38	39
40	41	42	43	44	45	46	47	48	49
50	51	52	53	54	55	56	57	58	59
60	61	62	63	64	65	66	67	68	69
70	71	72	73	74	75	76	77	78	79
80	81	82	83	84	85	86	87	88	89
90	91	92	93	94	95	96	97	98	99

Round down ⟵   Round up ⟶

Temperature

A **thermometer** is a tool that is used to measure temperature. The two scales of measuring temperature are Celsius (°C) and Fahrenheit (°F).

Description	Celsius	Fahrenheit
Boiling Point of Water	100°	212°
Normal Body Temperature	37°	98.6°
Room Temperature	20°	68°
Melting Point of Water (Ice)	0°	32°

Formulas for converting between Celsius and Fahrenheit.

Formula I — °C to °F
You must MULTIPLY by 9, DIVIDE by 5, then ADD 32

10°C to ? °F

10 x 9 = 90 | 90 ÷ 5 = 18 | 18 + 32 = 50°

Answer 10°C = 50°F

Formula II — °F to °C
You must SUBTRACT 32, MULTIPLY by 5, then DIVIDE by 9

212°F to ? °C

212 — 32 = 180 | 180 x 5 = 900 | 900 ÷ 9 = 100°

Answer 212°F to 100°C

Reciprocal

The reciprocal of a fraction can be made by interchanging the numerator and the denominator.

The reciprocal of $\dfrac{2}{3}$ is $\dfrac{3}{2}$

In the case of a whole number, think of it as having a denominator of 1:

The reciprocal of 7 is $\dfrac{7}{1}$ invert $\dfrac{1}{7}$

Reciprocal of a mixed number or mixed fraction. Change the mixed number into an improper fraction then invert (flip it over).

Reciprocal of 1 1/3 = change it into an improper fraction 4/3 then invert (flip it over) 3/4

Tessellation

A <u>tessellation</u> is created when a shape is used over and over again, covering a plane without any gaps or overlaps. Another word for a tessellation is a tiling.

a tessellation of triangles

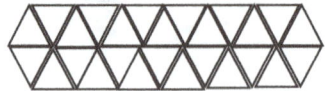

a tessellation of squares

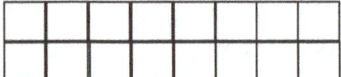

a tessellation of octagons

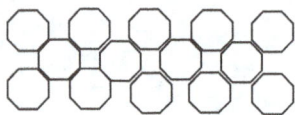

Outlier

The **outlier** is a value far away from most of the rest in a set of data.

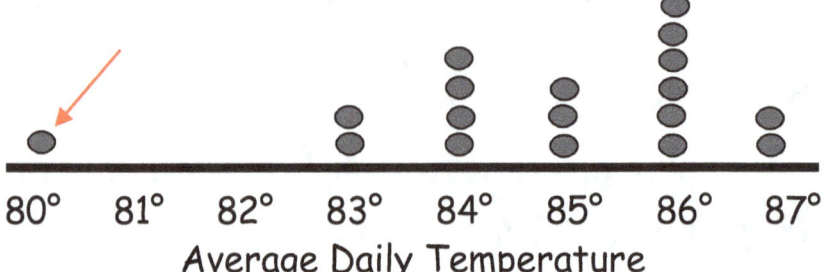

Expressions

An <u>expression</u> is a mathematical phrase that combines numbers, operation signs and sometimes variables. Expression don't use equal signs. For instance:

$$23 + 7$$
$$19 - 8$$
$$(34 + 6) - 2$$
$$9 \times 5$$

Variables

An expression may have a variable. A <u>variable</u> is a letter or symbol that represents an unknown member of a set. Here are some examples of possible variables: a, x, m, c, n, ★

$$n + 2 = 5$$

variable   operation   constant

$$16 - ★$$

constant   operation   variable

Descending and Ascending

Descending Order

Going down or decreasing in value. The following weights have been arranged in descending order.

545 lbs.	489 lbs.	302 lbs.	276 lbs.

biggest   to   smallest

5   4   3   2   1

Ascending Order

Going up or increasing in value. The following objects have been arranged in ascending order.

6 ft.	19 ft.	22 ft.	37 ft.	49 ft.

shortest   to   tallest

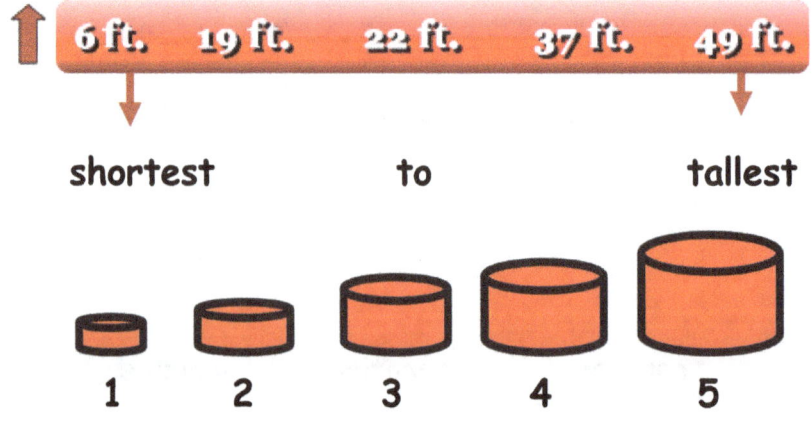

1   2   3   4   5

Congruent vs Similar

Congruent

Matching in exactly the same size and shape. It is the perfect pair. The congruent symbol is   . ≅

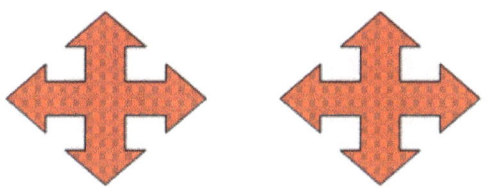

Similar

The same in shape but not in size. Two shapes are similar in figure if the corresponding angles are equal and all sides are enlarged or reduced by the same ratio.

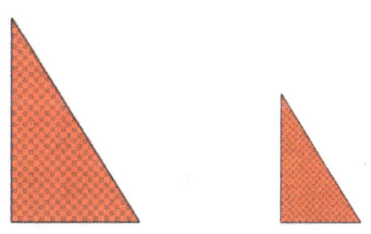

Squares

Square: To square a number just multiply it by itself. "Squared" is often written as a little 2.

this little two means "squared"

$$4^2 = 16$$

This says, "4 squared equals 16".

This is the "square" $3^2 = 9$ This is the "square root".

Perfect Squares
(Perfect Squares are the squares of whole numbers.)

Whole Numbers	Perfect Squares	Whole Numbers	Perfect Squares
1	1	11	121
2	4	12	144
3	9	13	169
4	16	14	196
5	25	15	225
6	36	16	256
7	49	17	289
8	64	18	324
9	81	19	361
10	100	20	400

Square Roots

A **square root** goes the other way: **3** squared is **9**, so the square root of **9** is a value that can be multiplied by itself to give the original number.

$$\sqrt{9} = 3$$

Square

3　　　9

Square Root

3 squared is 9, so a square root of 9 is 3

$\sqrt{9} = 3$　　　3^2　$3 \times 3 = 9$

expressed four ways

root index　radical symbol　radicand　solution/root

$^2\sqrt{81} = 9$

Square Roots
Example: The square root $\sqrt{}$ of 1 is 1 since $1^2 = 1$

$\sqrt{1} = 1$ since $1^2 = 1$	$\sqrt{64} = 8$ since $8^2 = 64$
$\sqrt{4} = 2$ since $2^2 = 4$	$\sqrt{81} = 9$ since $9^2 = 81$
$\sqrt{9} = 3$ since $3^2 = 9$	$\sqrt{100} = 10$ since $10^2 = 100$
$\sqrt{16} = 4$ since $4^2 = 16$	$\sqrt{121} = 11$ since $11^2 = 121$
$\sqrt{25} = 5$ since $5^2 = 25$	$\sqrt{144} = 12$ since $12^2 = 144$
$\sqrt{36} = 6$ since $6^2 = 36$	$\sqrt{169} = 13$ since $13^2 = 169$
$\sqrt{49} = 7$ since $7^2 = 49$	$\sqrt{196} = 14$ since $14^2 = 196$

Exponents

The <u>exponent</u> of a number says how many times to use the number in multiplication.

exponent
(also known as index or power)

base $\longrightarrow$ 8^2

In 2^4 the "4" says to use 2 four times in a multiplication.

So $2^4 = 2 \times 2 \times 2 \times 2 = 16$.

$$16 \;=\; 4 \;\times\; 4 \;=\; 4^2$$

value   factors   exponent
                                     form

Powers of Ten

Exponent Form	The Value	Word Form
10^1	10	Ten
10^2	100	Hundred
10^3	1,000	Thousand
10^4	10,000	Ten Thousand
10^5	100,000	Hundred Thousand
10^6	1,000,000	Million
10^7	10,000,000	Ten Million
10^8	100,000,000	Hundred Million
10^9	1,000,000,000	Billion

Expanded Form
10
10 x 10
10 x 10 x 10
10 x 10 x 10 x 10
10 x 10 x 10 x 10 x 10
10 x 10 x 10 x 10 x 10 x 10
10 x 10 x 10 x 10 x 10 x 10 x 10
10 x 10 x 10 x 10 x 10 x 10 x 10 x 10
10 x 10 x 10 x 10 x 10 x 10 x 10 x 10 x 10

Weighing I
60-40 Report Card Grading

PERCENT	TERM 60%	EXAM 40%	PERCENT	TERM 60%	EXAM 40%
100	60	40	69	41	28
99	59	40	68	41	27
98	59	39	67	40	27
97	58	38	66	40	26
96	58	38	65	39	26
95	57	38	64	38	26
94	56	38	63	38	25
93	56	37	62	37	25
92	55	37	61	37	24
91	55	36	60	36	24
90	54	36	59	35	24
89	53	36	58	35	23
88	53	35	57	34	23
87	52	35	56	34	22
86	52	34	55	33	22
85	51	34	54	32	22
84	50	34	53	32	21
83	50	33	52	31	21
82	49	33	51	31	20
81	49	32	50	30	20
80	48	32	49	29	20
79	47	32	48	29	19
78	47	31	47	28	19
77	46	31	46	28	18
76	46	30	45	27	18
75	45	30	44	26	18
74	44	30	43	26	17
73	44	29	42	25	17
72	43	29	41	25	16
71	43	28	40	24	16
70	42	28	39	23	16

Weighing II

PERCENT	TERM 60%	EXAM 40%		PERCENT	TERM 60%	EXAM 40%
38	23	15		19	11	8
37	22	15		18	11	7
36	22	14		17	10	7
35	21	14		16	10	6
34	20	14		15	9	6
33	20	13		14	8	6
32	19	13		13	8	5
31	19	12		12	7	5
30	18	12		11	7	4
29	17	12		10	6	4
28	17	11		9	5	4
27	16	11		8	5	3
26	16	10		7	4	3
25	15	10		6	4	2
24	14	10		5	3	2
23	14	9		4	2	2
22	13	9		3	2	1
21	13	8		2	1	1
20	12	8		1	1	0

CALCULATING G.P.A.
(using percentages)
- Firstly, add all final percentages for each subject area.
- Secondly, divide that by the number of subjects.
- Thirdly, multiply by 4
- Lastly, divide that answer by 100 to get the final G.P.A.

CALCULATING G.P.A.
(using ABC method)

A – 4.00
B – 3.00
C – 2.00
D – 1.00
F – 0.00

*Find the average.

GRADING SCALE

A. 100%-90%

B- 89%-71%

C - 70%-56%

D - 55%-45%

F - 44%-0%

Solid Figures

Solid figures are three-dimensional shapes. Many of the everyday objects, which are familiar, are solid shapes. Building blocks are often cubes or rectangular prisms.

Here are some easily recognized shapes.
1. Spheres are shaped like balls.
2. Cones are shaped like ice cream comes.
3. Cylinders are shaped like soda cans.

Solid Figures

cube	
rectangular prism	
pyramid	
sphere	
cylinder	
cone	

Shape	Faces	Edges	Vertices
cube	6	12	8
rectangular prism	6	12	8
Triangular prism	5	9	6
Square pyramid	5	8	5
Triangular pyramid	4	6	4
sphere	0	0	0
cylinder	2	0	0
cone	1	0	0

Triangles

A **triangle** is a plane figure with three straight sides and three angles.

Triangles Based of Sides		
Scalene	**Isosceles**	**Equilateral**
3mm 1mm 2mm	4in 4in 2in	5ft 5ft 5ft
Length of all sides are different	Length of two sides are equal	Length of all sides are equal

Triangles Based on Angles		
Acute	**Right**	**Obtuse**
Each angle is < 90° (less than 90)	One angle is = 90° (equal to 90)	One angle is > 90° (greater than 90)

Fractions

A <u>fraction</u> is a part of a whole. A number written with the bottom part (*the denominator*) telling you how many parts the whole is divided into, and the top part (*the numerator*) telling how many you have.

numerator → **1/4** ← denominator

proper fraction	improper fraction	mixed number
7/8	8/2	1¾

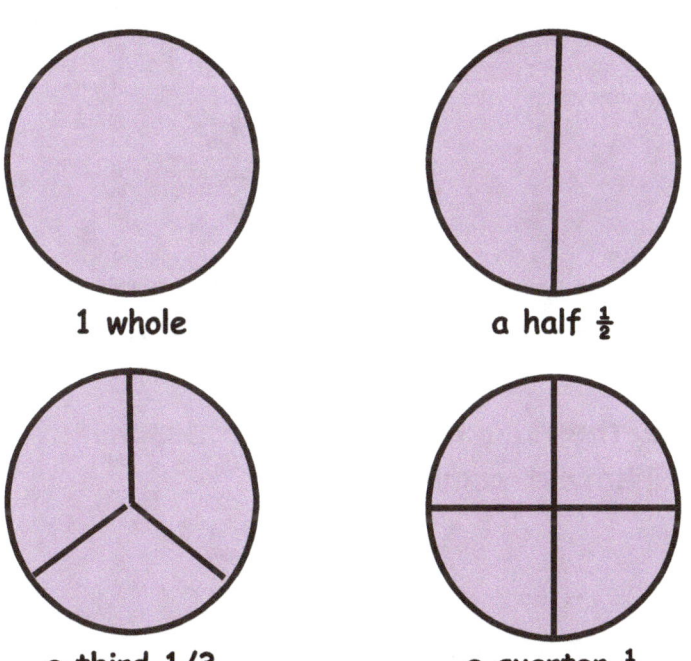

1 whole

a half ½

a third 1/3

a quarter ¼

Fractions

Equivalent fractions are fractions that have the same value. For instance:

$$\frac{2}{4} \quad \text{is equivalent to} \quad \frac{5}{10}$$

How to find the __equivalent of a fraction__?

1. by multiplying

$$\frac{1}{2} {}^{\times 2}_{\times 2} = \frac{2}{4}$$

2. by dividing

$$\frac{2}{4} {}^{\div 2}_{\div 2} = \frac{1}{2}$$

****

How to __simplify a fraction__?

Divide the numerator and the denominator by the greatest common factor.

$$\frac{6}{12} {}^{\div 6}_{\div 6} = \frac{1}{2}$$

Equivalent Fractions

1	2	3	4	5	6	7	8	9	10	11	12
1	2	3	4	5	6	7	8	9	10	11	12
2	4	6	8	10	12	14	16	18	20	22	24
3	6	9	12	15	18	21	24	27	30	33	36
4	8	12	16	20	24	28	32	36	40	44	48
5	10	15	20	25	30	35	40	45	50	55	60
6	12	18	24	30	36	42	48	54	60	66	72
7	14	21	28	35	42	49	56	63	70	77	84
8	16	24	32	40	48	56	64	72	80	88	96
9	18	27	36	45	54	63	72	81	90	99	108
10	20	30	40	50	60	70	80	90	100	110	120
11	22	33	44	55	66	77	88	99	110	121	132
12	24	36	48	60	72	84	96	108	120	132	144

original fractions

equivalent fractions

The multiplication chart hosts the equivalents.

0 zero	0/1	0/2	0/3	0/4	0/5	0/6	0/7
1	1/1	2/2	3/3	4/4	5/5	6/6	7/7
half	1/2	2/4	3/6	4/8	5/10	6/12	7/14
-	2/3	4/6	6/9	8/12	10/15	12/18	14/21
-	3/4	6/8	9/12	12/16	15/20	18/24	21/28
-	4/5	8/10	12/15	16/20	20/25	24/30	28/35 etc.

Fractions, Decimals, and Percent Chart

Fraction	Decimal	Percent	Think Money
$\frac{1}{4}$	0.25	25%	$0.25 quarter
$\frac{1}{2}$	0.50	50%	$0.50 half dollar
$\frac{3}{4}$	0.75	75%	$0.75
1/10	0.10	10%	$0.10 dime
1/20	0.05	5%	$0.05 nickel
1/100	0.01	1%	$0.01 penny
2/3	0.66	66 2/3%	
1/5	0.20	20%	
2/5	0.40	40%	
3/5	0.60	60%	
1/3	0.33	33 1/3%	
4/5	0.80	80%	
1/6	0.16	16 2/3%	
1/8	0.125	12 $\frac{1}{2}$ %	
3/8	0.375	37 $\frac{1}{2}$ %	
5/8	0.625	62 $\frac{1}{2}$ %	
7/8	0.875	87 $\frac{1}{2}$ %	
3/10	0.30	30%	
7/10	0.70	70%	
9/10	0.90	90%	
1/1	1 whole	100%	$1.00 a dollar

Conversion of Fractions, Decimals and Percent

From Percent to Decimal

Divide by 100, and remove the "%" sign. The easiest way to divide by 100 is to **move the decimal point 2 places to the left:**

$$75\% \qquad 0.7.5. \qquad 0.75$$

From Decimal to Percent

Multiply by 100, and add a "%" sign. The easiest way to multiply by 100 is to **move the decimal point 2 places to the right:**

$$0.125 \qquad 0.1.2.5. \qquad 12.5\%$$

From Fraction to Decimal

The easiest way is to divide the top number (numerator) by the bottom number (denominator).

$$2 \div 5 = 0.4$$

x 100 ÷ 1 and add a % sign

divide by 100, move decimal 2 to the left, zero fills gap

fraction	percent	decimal

put ÷ 100, remove % sign, and simplify

multiply by 100, move decimal 2 to the right, add a % sign

Geometrical and Numerical Patterns

Patterns are a sequence that repeats. There are two main types of patterns used in math: **numerical patterns** and **geometric patterns**.

A NUMERICAL PATTERN is a group of numbers that follow each other in a particular order. You must find the rule.

What comes next?   1, 3, 6, _____, _____. _____

The rule here is 1+2,+3,+4,+5+6 etc.

$$0+1=1$$
$$1+2=3$$
$$3+3=6$$
$$6+4=10$$
$$11+5=16$$
$$16+6=22$$

What comes next?   12, 24, 48, _____, _____, ___

The rule here is x2

$$12\times2=24$$
$$24\times2=48$$
$$48\times2=96$$
$$96\times2=192$$
$$192\times2=384$$

*Hint – Start with what is the difference between the two numbers that are next to each other.

A GEOMETRIC PATTERN uses a variation of repeated objects and shapes.

i. **Square Number Sequence** is the result of multiplying a number by itself. Formula:

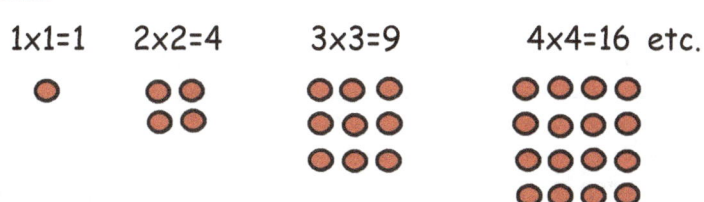

1x1=1   2x2=4   3x3=9   4x4=16 etc.

1, 4, 9, 16, 25, 36, 49, 64, 81, 100, 121, 144, etc.

ii. **Triangular Number Sequence** is the result of adding another row of dots and counting all the dots you can find. Formula:

1   1+2=3   1+2+3=6   1+2+3+4=10

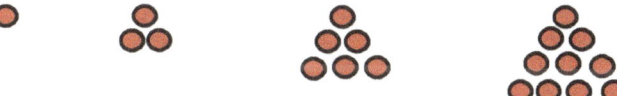

1, 3, 6, 10, 15, 21, 28, 36, 45, 55, 66, 78, 91, etc.

iii.**Oblong Number Sequence** is the result of multiplying consecutive numbers. Formula:

1x2=2   2x3=6   3x4=12   4x5=20 etc.

2, 6, 12, 20, 30, 42, 56, 72, 90, 110, 132, etc.

Number Prefixes

prefix	meaning	example
mono	1	monochord
bi	2	bicycle
tri	3	tricycle
quad	4	quadruplet
petra	5	pentagon
poly/multi	many	polygon
semi	half	semicircle
uni	one	unit

Terms for Multiple Babies

	Multiple offspring
1	single
2	twins
3	triplets
4	quadruplets
5	quintuplets
6	sextuplets
7	septuplets
8	octuplets
9	couplets
10	decuplets

Adding Fractions

ADDING LIKE FRACTIONS

$$\frac{3}{8} + \frac{4}{8}$$

Step 1: Look for like denominators. If the denominators are alike then add ONLY the numerators. Mirror the denominator.

$$\frac{3}{8} + \frac{4}{8} = \frac{7}{8}$$

SUBTRACTING LIKE FRACTIONS

$$\frac{9}{12} - \frac{6}{12}$$

Step 1: Look for like denominators. If the denominators are alike then add ONLY the numerators. Mirror the denominator.

$$\frac{9}{12} - \frac{6}{12} = \frac{3}{12}$$

Step 2 – Simplify the answer if possible by finding the GCF (Greatest Common Factor).

3 - (1, 3)
12 - (1, 2, 3, 4, 6, 12)   GCF = 3

$$\frac{3}{12} \div \frac{3}{3} = \frac{1}{4}$$

ADDING UNLIKE FRACTIONS

$$\frac{2}{5} + \frac{1}{3}$$

Step 1: Find the LCM (Least Common Multiple) of the denominators.

$$5 - 5, 10, 15, 20, 25, 30, 35 \text{ etc.}$$
$$3 - 3, 6, 9, 12, 15, 18, 21, 24 \text{ etc.}$$
$$\text{LCM} = 15$$

Step 2: Create equivalent fractions using the LCM as the new denominator.

(Ask yourself: 5 times what equals 15. The ans. is 3. And 3 times what equals 15? The ans. is 5)

$$\frac{2}{5} \times \frac{3}{3} \qquad \frac{1}{3} \times \frac{5}{5}$$

Step 3: So then, multiply the answer (3 and 5) by the numerator and denominators.

$$\frac{2}{5} \times \frac{3}{3} = \frac{6}{15} \qquad \frac{1}{3} \times \frac{5}{5} = \frac{5}{15}$$

Step 4: Now that both denominators are the same, you can add your new like fractions.

$$\frac{6}{15} + \frac{5}{15} = \frac{11}{15}$$

Subtracting Fractions

SUBTRACTING MIXED FRACTIONS

$$2 \frac{3}{4} \quad - \quad 1 \frac{1}{7}$$

Step 1 – Convert the mixed numbers into improper fractions.

$$2 \frac{3}{4} \quad - \quad 1 \frac{1}{7} \quad = \quad \frac{11}{4} \quad - \quad \frac{8}{7}$$

$$4 \times 2 + 3 = 11 \qquad 1 \times 7 + 1 = 8$$

Step 2 - Find the LCM of the denominators.

4 – 4, 8, 12, 16, 20, 24, 28 etc.
7 – 7, 14, 21, 28, 35, 42 etc.
LCM = 28

Step 2: Create equivalent fractions using the LCM as the new denominator.

(Ask yourself: 4 times what equals 28. The ans. is 7. And 7 times what equals 28? The ans. is 4)

$$\frac{11}{4} \times \frac{7}{7} \qquad 8 \times \frac{4}{4}$$
$$\qquad\qquad\quad 7$$

Step 3: So then, multiply the answer (7 and 4) by the numerator and denominators.

$$\frac{11}{4} \times \frac{7}{7} = \frac{77}{28} \qquad \frac{8}{7} \times \frac{4}{4} = \frac{32}{28}$$

Step 4: Now that both denominators are the same, you can subtract your new like fractions

$$\frac{77}{28} - \frac{32}{28} = \frac{45}{28}$$

Step 5 – Change the improper fraction to a mixed number by dividing it.

$$28 \div 45 = 1 \text{ R}17$$

$$1 \ \frac{17}{28}$$

*Adding and Subtracting Fractions uses the same steps.

Multiplying Fractions

MULTIPLYING FRACTIONS

$$\frac{1}{4} \quad \times \quad \frac{2}{3}$$

Step 1 - Multiply the numerators, then the denominators.

$$\frac{1}{4} \quad \times \quad \frac{2}{3} \quad = \quad \frac{2}{12}$$

Step 2 – Simplify the answer if possible by finding the GCF (Greatest Common Factor).

2 – (1, 2)
12 – (1, 2, 3, 4, 6, 12)

GCF = 2

$$\frac{2}{12} \quad \div \quad \frac{2}{2} \quad = \quad \frac{1}{6}$$

MULTIPLYING MIXED NUMBERS

$$3 \times \frac{3}{4}$$

Step 1 – Find the reciprocal of the whole number.

$$3 = \frac{3}{1}$$

Step 2 – Multiply the fractions.

$$\frac{3}{1} \times \frac{3}{4} = \frac{9}{4}$$

Step 3 – Change the improper fraction to a mixed number by dividing it.

$$4 \div 9 = 2 \; r \; 1$$

$$2 \frac{1}{4}$$

MULTIPLYING MIXED NUMBERS

$$4 \frac{1}{5} \times 2 \frac{1}{3}$$

Step 1 – Convert the mixed numbers into improper fractions.

$$4 \frac{1}{5} \times 2 \frac{1}{3} = \frac{21}{5} \times \frac{7}{3}$$

4 x 5 + 1 = 21   2 x 3 + 1 = 7

Step 2 – Multiply the two improper fractions.

$$\frac{21}{5} \times \frac{7}{3} = \frac{147}{15}$$

Step 3 – Simplify the answer if possible by finding the GCF (Greatest Common Factor).

147 - (1, 3, 7, 21, 49, 147)
15 - (1, 3, 5, 15)

GCF = 3

$$\frac{147}{15} \div \frac{3}{3} = \frac{49}{5}$$

Step 4 – Convert the improper fraction into a mixed number by dividing.

5 ÷ 49 = 9 r 4

9   4/5

Dividing Fractions

$$\frac{2}{3} \div \frac{1}{4}$$

Step 1 – Find the reciprocal of the second fraction.

$$\frac{1}{4} \quad \text{becomes} \quad \frac{4}{1}$$

Step 2 – Multiply the numerators, then the denominators.

$$\frac{2}{3} \times \frac{4}{1} = \frac{8}{3}$$

Step3 – Convert the improper fraction into a mixed number by dividing.

$$3 \div 8 = 2 \text{ r } 2$$

$$2 \quad 2/3$$

Vocabulary

1. **absolute value**—How far a number is from 0 on a number line, either side of 0. The absolute value of a number is never negative.

2. **annually**—happening once a year

3. **anticlockwise**—Turning the opposite way from the hands on a clock.

4. **apex**—the highest point, the top of a shape (i.e. pointy end of a cone)

5. **arc**—a section of a curve, a part of a circle

6. **array**—An arrangement of objects or numbers in columns or rows.

7. **ascending order**—Going upward or increasing in value.

8. **axis**—The lines that make a graph's framework.

9. **clockwise**—The direction in which the hands of a clock normally travel.

10. **cluster**—Numbers which tend to crown around a particular point in a set of numbers.

11. **compass** - an instrument which shows direction

12. **congruent** - Having exactly the size and shape.

13. **data**—A collection of facts, numbers, measurements, or symbols.

14. **debit**—The amount of money taken out of an account.

15. **decrease**—to get smaller in size or number

16. **degree**—In geometry, a degree is a unit used for measuring angles.

17. **descending order**—Going down or decreasing in value.

18. **diagonal**—Something that is slanting.

19. **digit**—Numerals 0 to 9 are called digits. They are used to make other numbers like 19 & 463.

20. **dozen**—a group or set of 12

21. **elapsed time**—The time that has passed between the start of an activity and the end of that activity.

22. **estimate**—to make a rough calculation

23. **evaluate**—to find the numeric value for something or "to work it out"

24. **factor tree**—A diagram showing prime factors.

25. **finite**—able to be counted, limited

26. **fortnight**—two weeks or fourteen days

27. **graph** - a diagram showing the relationship between variable quantities, typically of two variables.

28. **horizontal**—Being positioned from side to side or parallel to the ground

29. **hypotenuse**—In a right triangle, the side opposite the right angle; which is the longest side.

30. **increase**—to get larger in size or number

31. **infinite**—An unlimited or endless number

32. **intervals**—The distance between one number and the next on the scale of a graph.

33. **km/h**—kilometers per hour (speed)

34. **leap year**—occurs every 4 years and has 366 days

35. **mph**—miles per hour (speed)

36. **oblong**—a type of rectangle

37. **operations**—There are four basic operations in arithmetic i.e. add, subtract, multiply and divide.

38. **order**—arrange according to value, size, or amount

39. **ordered pairs**—a pair of numbers where order is important i.e. (5, 6) is different to (6, 5)

40. **outcome**—one of the possible results of a probability experiment

41. **pattern**—repeated design or occurring sequence

42. **population**— whole set of individuals, items, or data from which a statistical sample is drawn

43. **plane shapes (2D shapes)** —flat shapes having only two dimensions

44. **profit**— occurs when and item is sold for more than it cost to buy

45. **prioritize**- To put events in order of importance.

46. **prism**—A 3-D shape with two parallel faces that are polygons, and the same in shape and size.

47. **probability** -- The likelihood, or chance, of an event happening.

48. **protractor**—A device that has the form of a half circle and that is used for drawing and measuring angles.

49. **quadrant**—a quarter of a circle or its circumference

50. **quantity**—amount or number of something

51. **quarterly**—relating to quarters, made up of four parts; four times a year (every three months)

52. **regroup**—used to assist when trading or carrying in addition or subtraction

53. **reoccurring decimal**— a decimal which has repeated digits or patters or a digit i.e. 1/3 = 0.3333333

54. **revolution**—one complete turn through 360 degrees

55. **row**—items arranged in a horizontal line

56. **scientific notation**—is a way of writing very large or very small numbers using a number between 1 and 10 multiplied by a power of 10.

57. **sequence**-an ordered set of numbers, shapes, or other mathematical objects arranged according to a rule

58. **simplify**—To write something in the simplest or shortest form.

59. **tangent**— a straight line toughing a curve once at a given point

60. **unit**—another name for one;

61. **unlikely**—will probably not happen

62. **value**—numerical worth or amount

63. **venn-diagram**—A Venn diagram is used to sort things into groups or sets. It shows how sets are related.

64. **vertical**– Being positioned up and down or going straight up.

65. **whole**—all, everything, total amount, all parts

66. **x-axis**—a horizontal axis on a graph

67. **y—axis** —a vertical axis on a graph

68. **zero**—used as a place holder; meaning nothing, none, nil, naught

www.ingramcontent.com/pod-product-compliance
Lightning Source LLC
Chambersburg PA
CBHW070316240526
45467CB00045B/497